W0254135

For Noah – M.B.
For Mei Mei – A.M.

BLOOMSBURY CHILDREN'S BOOKS
Bloomsbury Publishing Plc
50 Bedford Square, London, WC1B 3DP, UK
29 Earlsfort Terrace, Dublin 2, Ireland

BLOOMSBURY, BLOOMSBURY CHILDREN'S BOOKS and the Diana logo are trademarks of Bloomsbury Publishing Plc

First published in Great Britain 2024 by Bloomsbury Publishing Plc

A catalogue record for this book is available from the British Library

ISBN: PB: 978-1-5266-4708-5; eBook: 978-1-5266-6908-7
2 4 6 8 10 9 7 5 3 1

Printed in China by RR Donnelley Asia Printing Solutions LTD, Dongguan City, Guangdong Province

To find out more about our authors and books visit www.bloomsbury.com and sign up for our newsletters

Moira Butterfield

Adam Ming

DOES A BEAR WASH ITS HAIR?

BLOOMSBURY
CHILDREN'S BOOKS
LONDON OXFORD NEW YORK NEW DELHI SYDNEY

We all have things we do each day
to live our lives the human way ...

keeping clean

getting dressed

making friends

eating meals

tidying up

learning

brushing teeth

going to sleep

Animals do the same things, too,
but differently from me and you.
How do they **eat** and **wash** and **poop**?
Let's meet a few and get **the scoop**!

KEEPING CLEAN

Brown bears lick their fur to keep clean.
They **slurp** up dirt and dust and bugs.
For them, a **squishy** bug is sweet,
so when they **wash**, they get a treat.

Keeping their fur clean helps bears to keep warm and dry. They sometimes swallow loose hairs by mistake, but they can just poo them out later.

I KEEP CLEAN, TOO

Cleaner wrasse run cleaning zones
for bigger fish on coral reefs.
They'll nibble off bugs and flaking scales
from fins and tummies, gills and tails.

Even scary hunter fish like this green moray eel will happily visit a coral reef cleaning station to let cleaner wrasse tidy them up.

GETTING DRESSED

Decorator crabs dress up
with ocean scraps of this and that.
Seaweed, coral, anemones too –
any **feathery bits** will do!

The crabs have tiny bristles on their shells to stick decorations on. The bristles work like the Velcro that you might have on clothes and shoes.

It sits down **silently** on the seafloor, not looking **crabby** anymore.

Just ... act ... natural ...

In its **disguise**, it hopes to **hide** from **enemies** with **beady eyes**.

Phew! I'm safe!

These little crabs don't wear decorations to be fashionable. The items are used for camouflage, disguising the crab on the seabed to fool hungry predators.

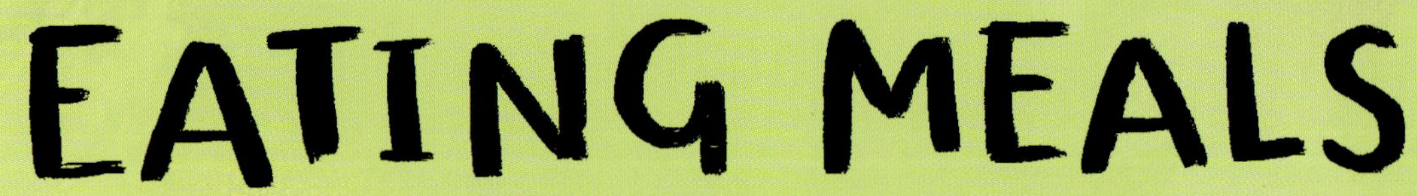

EATING MEALS

An **anteater** can smell out **ants** and rip nests open with its claws.

An anteater needs to find and eat roughly 35,000 ants a day around their South American homes. Imagine if you had to spend all day searching for that many teeny-tiny bits of food to fill your stomach!

Then its sticky **tongue** flicks out
to lick up all the ants about.
Ants will **sting** the anteater's tongue,
so it needs to be **quick** or it'll get stung!

Like us, an anteater eats every day, but some animals need to eat far less often. Galápagos tortoises can go a whole year without a meal! This is because they can store food and water in their bodies for a long time.

POOING

A **sloth** will climb down from its tree to take a **big poo** once a week. It poos out a third of its body weight. **Eek**!

It's dangerous for sloths to poo out in the open too often. Once they've done their pile of poos, they bury it so their enemies won't find out where they live.

I POO, TOO

Blue whales do giant glowing poos.
They're coloured orange, pink or green –
the **brightest poo** you've ever seen!

Blue whales eat pinky-red krill, which is why their poo is so brightly coloured. What you eat can change the colour of your poo, too. For example, eating beetroot might make your poo red.

Rabbits do the same **poo twice**.
The first one's **soft** and **small** and **round**.
Then they eat the ball back up,
and poo a **hard one** on the ground!

Eating the first poo doesn't harm the rabbit. The poo is just partly-chewed food and is actually full of nutrients (goodness) for the rabbit.

LEARNING

Meerkat pups have **hunting lessons** where they learn to bite off a scorpion's sting.

The scorpion makes a **yummy meal** but its venom is a **dangerous** thing!

Baby meerkats first practise with a dead scorpion. Then they get a live one with its stinger already removed. Soon, they're ready to tackle the real deal!

Before a **meerkat** eats a millipede,
it **rolls** it in sand to rub off its **poison**.
That's something the babies have to learn,
or the poison will make their insides **burn**!

Around 40 meerkats live together in burrows in southern Africa. The big ones teach skills to the small ones, just as you might learn from your teachers or other grown-ups in your life.

TIDYING UP

A **pregnant fox** will make a den and **scrape** away the muck and mess. She'll **clean** the tunnels thoroughly before she has her family.

Animals who live in burrows will scrape out rubbish, such as dead plants and loose soil, to make sure their home is a healthy place to be. We keep our homes clean for the same reason.

I TIDY, TOO

New Caledonian crows use sticks to **fish out** tasty grubs to eat. Then they tidy their **tools** away to store them for another day.

WORKING TOGETHER

Wild wolf packs show us how an animal group can **get along**. Their **sounds** and **body signals** send information to their friends.

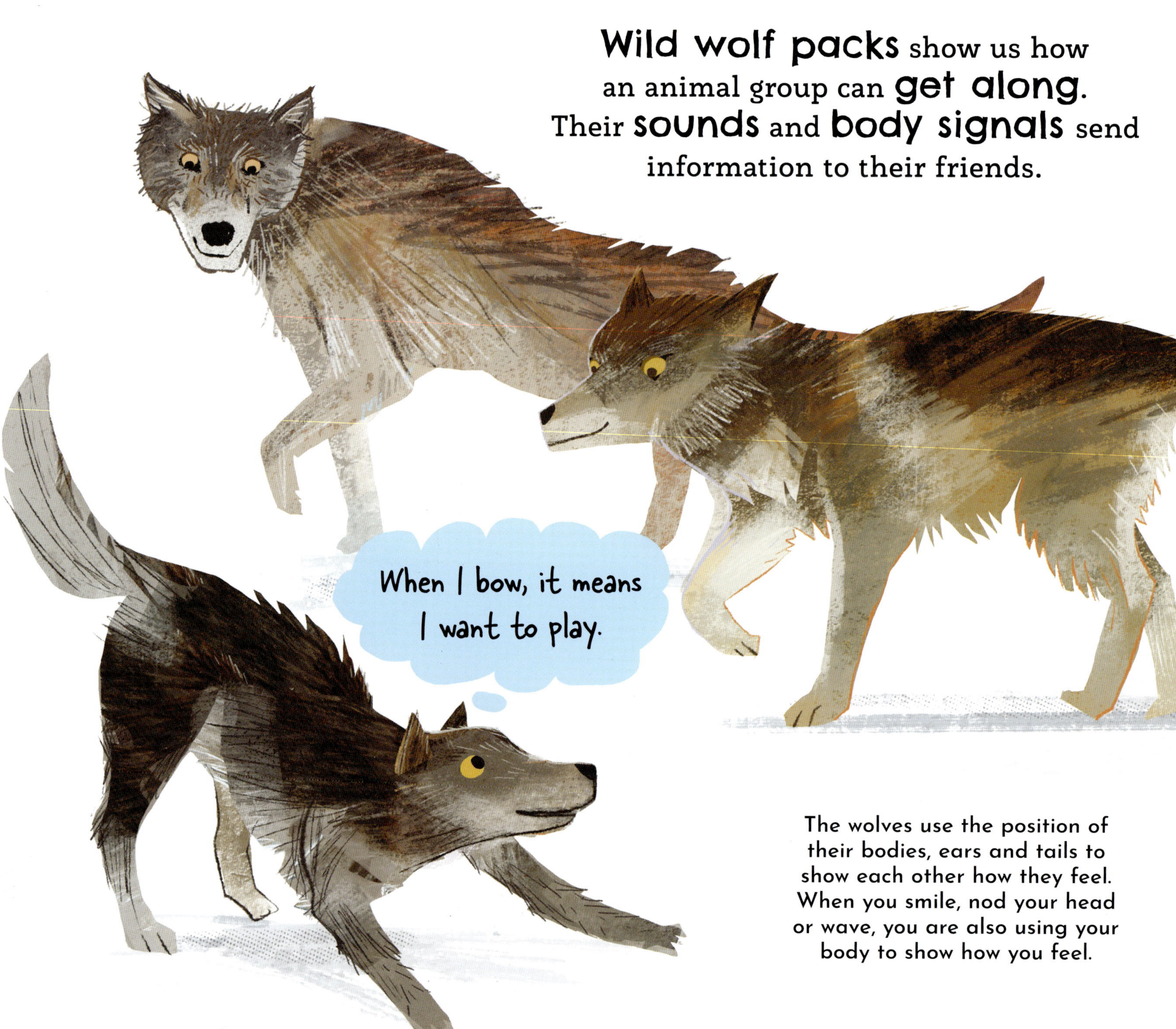

The wolves use the position of their bodies, ears and tails to show each other how they feel. When you smile, nod your head or wave, you are also using your body to show how you feel.

Sometimes a wolf pack **howls** at night, like **singing together** in a choir. The little wolves learn howling songs by **copying** the older ones.

By howling together, the pack is saying "We're great! We'll be strong together when we go hunting." It's like a human sports team working together and giving each other high-fives before a match.

MAKING FRIENDS

Some animals who live in groups
have **special friends** they like the best.
They'll **help** their buddies through the day
and **never wander** far away.

Cows are very social animals, and have best friends in their herd. They graze together, lick each other and even sleep up close to their pals. These relationships relieve stress and bring them happiness, just like our friendships do!

I MAKE FRIENDS, TOO

Alpaca friends like **rubbing noses** with each other on the farm. They'll also **guard** their friends from harm.

Alpacas will also make friends with other farm animals such as goats, horses and dogs. It's a great idea to make all kinds of different friends, just like they do!

BRUSHING TEETH

Crocodiles have a clever trick to keep their **long teeth** squeaky clean. They stretch their big jaws **open wide** and **plover birds** hop right inside!

There's a feast between my teeth!

This unusual tooth-cleaning happens on the Nile River in Egypt, where crocodiles and Egyptian plover birds live. The crocodiles won't eat the birds though, because they are doing them a favour.

These birds aren't just being kind.
They **eat** everything they find!

Most creatures don't do anything to clean their teeth but a few clever animals, such as gorillas and macaques, use twigs or bits of grass. We humans are the top tooth-cleaners in the animal kingdom!

GOING TO SLEEP

Every night **orangutans** build a **leafy nest** up high.
They bend branches and weave leaves
to make a **cradle** in the **sky**.

Wild orangutans make nests up high in the rainforest trees of Sumatra and Borneo. While they sleep, they can stay safe in their nest, hidden from hungry jungle predators.

Young orangutans learn from **their mum** how to make a **treetop nest.**

Zzzzzzz ...

It keeps them safe and cosy and is perfect for a **good night's rest.**

Orangutans sleep around eight hours a night. They even dream – just like you!

Now when you **wash** or **eat** or **sleep**
and do the daily things you do,
think of all the **animals**
who live in ways a **bit like you**!

Clever, quick, big or small.

Nosy, noisy, short or tall.

Let's learn what we can about them all!

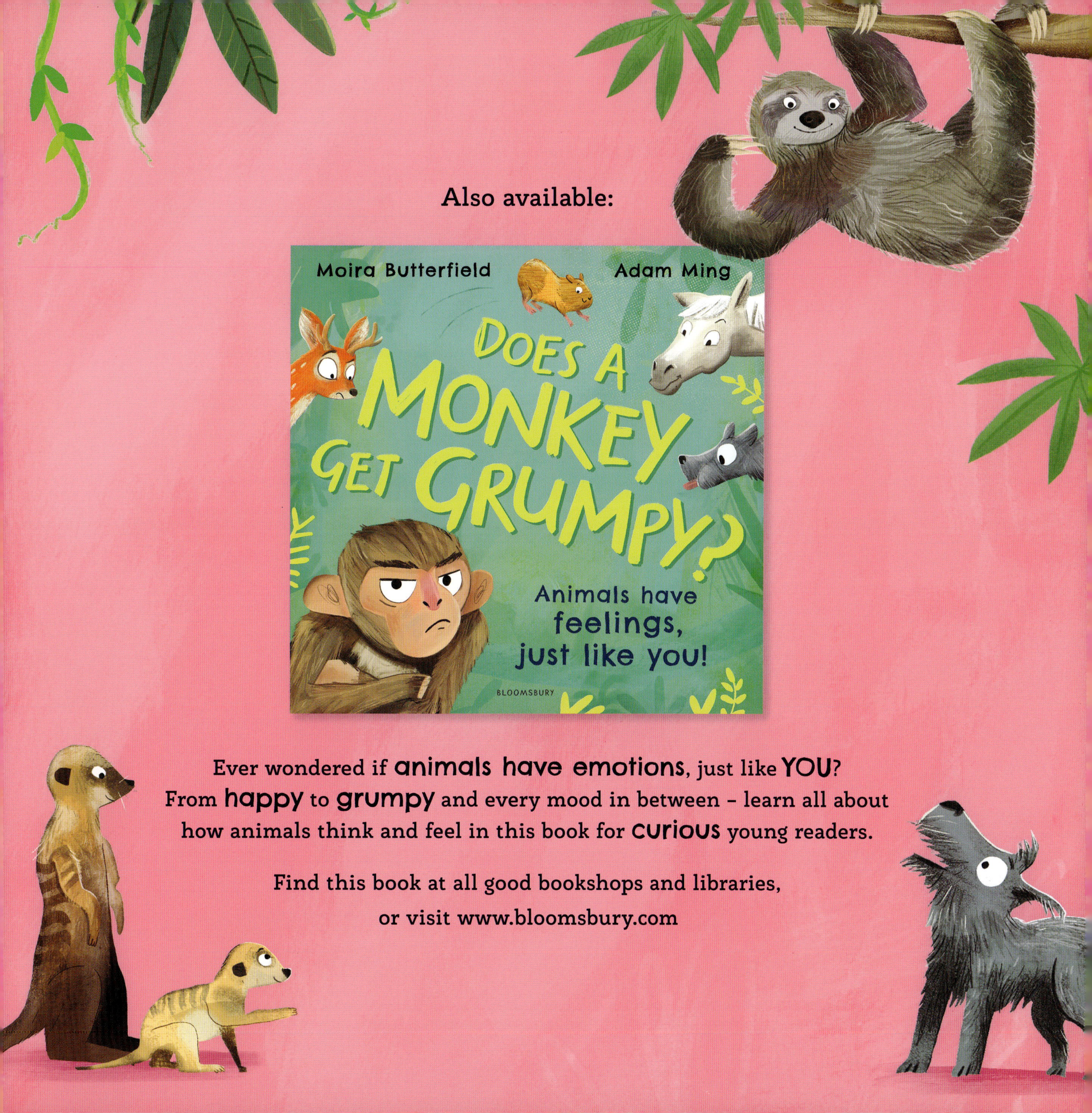

Also available:

Ever wondered if **animals have emotions**, just like **YOU**?
From **happy** to **grumpy** and every mood in between – learn all about how animals think and feel in this book for **curious** young readers.

Find this book at all good bookshops and libraries,
or visit www.bloomsbury.com